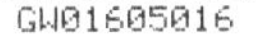

how2become

London Bus Driver
The Insider's Guide

Orders: Please contact How2become Ltd, Suite 2, 50 Churchill Square Business Centre, Kings Hill, Kent ME19 4YU.

Telephone: (44) 0845 643 1299 - Lines are open Monday to Friday 9am until 5pm. Fax: (44) 01732 525965. You can also order via the e mail address info@how2become.co.uk.

ISBN: 978-1-907558-20-7

First published 2010

Typeset for How2become Ltd by Good Golly Design, Canada, goodgolly.ca

Printed in Great Britain for How2become Ltd by: CMP (uk) Limited, Poole, Dorset.

CONTENTS

INTRODUCTION

Are you seeking a rewarding career with considerable potential? If so, then becoming a London bus driver might just be one of the best options available to you. Buses have an incredibly long history in the city of London, actually dating back to the middle of the 1800s. Since that time, the image of the ubiquitous double-decker bus purring through the London city streets has become ingrained in the minds of Britons and people from nations all over the world.

Why might becoming a London bus driver be a rewarding career choice? First, it offers you the chance to enjoy a lucrative career. Of course, the benefits of the job do not simply stop with a decent salary. You will also find that this career choice gives you the ability to meet interesting people every day, enjoy the challenge of navigating through the city of London, explore this world-famous city to your heart's content and more.

Of course, one of the most rewarding aspects of being a bus driver for the city of London is the fact that you will be providing an essential service to those in need. Whether you choose to operate a city bus, or want to work with a charter service offering help to visitors from around the world, you will be able to provide assistance to travellers in their time of need.

Finally, you will also find that becoming a London bus driver gives you the ability to choose from quite a few different companies. There are far more than a handful of coach companies operating in the city of London, and they can offer very different pay and benefit packages for their drivers. This ensures that you are able to find a rewarding job, with a company that you actually enjoy working for.

This book will give you all the information you need about becoming a bus driver in the city of London. From the industry's birth to required training, from potential companies for employment to valuable links and vital resources, this book is designed to help you through the entire process and ensure your success.

CHAPTER 1

THE HISTORY OF THE LONDON BUS

The city of London is synonymous with many things. Some people might think of the Thames, flowing steadily through the city on its way to the sea. Other people might think of the London Eye, the Tower of London, Paddington Station, the Queen Mother or even world-class cuisine. However, if you asked most people in the world what they first thought of when they heard the city's name, the answer would likely be "buses".

Bus driving has an incredibly long and rich history in the city. In fact, the rear entrance double-decker bus has become one of the internationally recognised symbols of the city. While similar buses are used in other world-class cities, it is inextricably tied to the city of London. In this chapter, we'll learn a bit more about the history of bus driving in London and how the industry has changed over the years. Before you consider skipping this chapter, consider this.

During the interview you may possible get asked a question about your knowledge of London and in particular the history of the London Bus. By reading this chapter you will be increasing your knowledge of the history that surrounds your chosen career.

IN THE BEGINNING WAS THE OMNIBUS

The year was 1829 and George Shillibeer had just started a new travel service in the city of London. His omnibus service made the trip from Paddington into the heart of London, offering travellers a welcome break from walking. The omnibus service was actually a horse-drawn bus.

Thomas Tilling got into the act around 1850, with his horse-drawn bus service. These new services were enormously popular with locals and visitors alike. In fact, the new service seemed so popular that others soon got into the act and, shortly, London had all manner of bus services plying the roadways.

With the explosion of service providers and the multiplying of buses on the city's roadways, the need for greater organisation was seen. To help combat problems and congestion stemming from an unregulated industry, the London General Omnibus Company (LGOC) was formed in 1855.

The job of the LGOC was to join together the various companies under one upper-level firm, which would provide organisation for all of the various firms. The horse-drawn omnibus remained in use in the city until 1911, when the last LGOC bus made its trip through the city (private operators continued using horses until 1914, though).

MOTORISED BUSES COME TO LONDON

Interestingly, the motor-powered bus actually started plying the city roadways in 1902. At first, the LGOC outsourced their buses, but beginning in 1909, they actually began manufacturing their own buses. 1904 saw Thomas Tilling Company begin their first motorised bus service through the city, as well.

In 1909, the LGOC and Thomas Tilling combined their resources into a larger conglomerate, and the same year saw steam-powered buses hit the city's streets via Thomas Clarkson and National Steam Car Company. Steam bus service only lasted for another 10 years, though, as National withdrew from the industry in 1919, after reaching an agreement with the LGOC.

The Underground Group purchased the LGOC in 1912, bringing the smaller company under the control of the same entity that owned most of the London Underground system. This group changed its name again in 1933, becoming the London Passenger Transport Board, to better reflect the various enterprises under the umbrella name.

LGOC still technically existed as a company, though as a subservient entity, and in 1933, its name changed to London Transport (from which stemmed the traditional red bus).

After this, the LPTB began serving passengers within a 30-mile radius of central London, and amalgamated other bus companies, such as parts of the Tilling Group, as well as Green Line Coaches. The LPTB continued until 1948, when it was renamed the London Transport Executive.

In 1963, it became the London Transport Board, and in 1970, it became the Greater London Council. 1984 marked the change to the name London Regional Transit. However, today, it is part of the Greater London Authority, under the name Transport for London.

THE RISE OF AN ICON – BUS VEHICLES IN LONDON

Interestingly, in a departure from almost any other city in the UK (or the world, for that matter), London bus designs were created specifically for the city environs. While other cities made use of standardised bus designs that can still be seen throughout the world, the city of London went its own way.

The company responsible for designing most of the early London buses was the Associated Equipment Company, a subsidiary of LGOC, founded in the early 1900s. This company continued to design and manufacture buses specifically for use in the city of London, and the company even sold buses to transport organisations that were not part of the LGOC, but operated in the general area.

The final bus designed and built for use in the city of London by AEC was the Routemaster. It was produced from 1956 through 1968 and made such an impact on the city and the residents thereof that every bus designed for specific use within the city from that point on has simply been a modernisation of the same design. This was the iconic, red, double-decker, rear entrance bus for which the city would become famous.

During the 1980s and 90s, midsize buses and even minibuses began to operate throughout the city, but they were unable make a dent in the popularity of the original bus style, and soon fell out of favour, though small, niche operations can still be found throughout the city that utilise these bus types.

In 2000, new double-decker buses were added to the fleet, which included a group of "bendy buses" or articulated buses.

These were not terribly popular with residents or visitors and changes were made to the TfL in 2008, when newly elected Mayor Boris Johnson announced a competition to design a new bus specifically for the city of London.

2009 saw the full removal of the unpopular bendy buses, though that withdrawal would be staggered, as each bus reached the end of its five-year operational contract. This allowed the TfL to enjoy somewhat reduced costs for bus replacement, as well as keeping the number of operating buses in the city relatively stable.

The winning design of the new city bus was announced in 2010. The new design, which will not actually see use on the city streets until 2012, will feature three different doors, two staircases and an open "hop on/hop off" platform for passengers to use.

THE DOOM OF THE RED DOUBLE-DECKER?

Some of those opposed to the new bus design program were concerned that the iconic double-decker bus would be forced out of the city's transportation scheme. This was due to a renewed emphasis on lowered floor buses, and on more environmentally friendly vehicles. However, modern buses operating within the city can have either one or two decks, which means that the double-decker bus will continue in use.

However, the Routemaster is no longer used for most of the city's public transport needs. In fact, the iconic bus is now restricted to just two routes within the city, called heritage routes. These buses are primarily used for tourists travelling through the city who wish to enjoy the "London bus" experience as it was in its heyday.

A NOTE ON HERITAGE ROUTES

There are two heritage routes within the city of London that still employ the older style Routemaster buses. These are heritage route 15, which runs from Trafalgar Square to Tower Hill, and heritage route 9, which makes the trip from Royal Albert Hall to Aldwych. As a note, these routes are part of the London Buses network. They utilise the same fares as the rest of the London Buses network, as well.

As a further note, Routemaster buses are not accessible to the handicapped (particularly those bound to a wheelchair). Therefore, standard lowered floor buses also operate on these two heritage routes, offering disabled passengers the means to enjoy the trips, as well. This ensures that all passengers have equal transport opportunities throughout the area.

NIGHT BUSES AND TOUR BUSES

Night transport via bus has long been available throughout London. In fact, the first night time routes began back in 1913. Today, night buses remain an integral part of the city's transportation network and are part of the London Bus network.

Many night routes are denoted by an N in front of the route number. For instance, the 9 route is a daytime route, while the N9 is a night route. However, in recent years, some bus routes have changed to a 24-hour system. These maintain the same numbering system as used for daytime routes, with no N before the route number.

Other routes have changed in different ways. For instance, originally, night routes were more expensive for travellers, due to their greater range of service. However, in recent days, some of these previously higher fare routes have reduced their prices to normal daytime fares to encourage passengers and make the routes more profitable (higher fares were discouraging riders on some routes).

In addition, all night buses are standard red buses, and the night service has increased in the number of passengers by as much as 80% over the early years of the 21st century. Much of this change and growth is due to efforts on the part of former Mayor Ken Livingstone.

Tour buses are also an integral part of the city's transportation network. The majority of buses operating for tour purposes are open top buses, though there are some closed top buses on these routes, as well. There are several tour companies that operate routes with open toppers, including the following:

- York Pullman
- Transdev York (York City Sightseeing)
- City Sightseeing
- Lothian Buses
- The Original Tour
- London Pride Sightseeing
- The Big Bus Company
- Ensignbus
- Bluestar
- Guide Friday

In addition to being used for touring and sightseeing purposes, open toppers are also used for several seasonal routes in and around the city of London. For instance, Quantock Motor Services operates route 300, which runs from Somerset through Pool Harbour, and then down to Lynmouth, as well as route 400, which begins in Minehead and runs through Exmoor.

Other companies offering seasonal bus service throughout the UK with open toppers include First Devon and Cornwall, Wilts & Dorset, East Yorkshire Motor Services, Stagecoach Devon and numerous others, as well.

CHAPTER 2

HOW TO APPLY TO BECOME A LONDON BUS DRIVER

If you are interested in becoming a London bus driver, then you certainly need to know how to get the process started. What should you know? What are the requirements to even apply with a bus company? This chapter will explore the topic of getting started in some depth, ensuring that you have all the information you need to get started as quickly and efficiently as possible.

UNDERSTANDING THE JOB – THE NATURE OF BUS DRIVING

Before we delve into the topic of how to get started as a bus driver, a word about the nature of the position is needed. What might you need to know? Understanding the work environment is very important, as driving a bus is a far different experience from handling a passenger car or lorry in the city.

In addition to being able to adequately handle a large bus in traffic and other situations, you will also need to be able to handle your passengers with equanimity. Be able to provide 'excellent' customer service is paramount to the job.

The driving environment is a unique one. While those working in an office environment will be able to get up and "stretch their legs" throughout the workday, bus drivers do not have this luxury. Being a bus driver requires that you are able to sit for long periods behind the wheel. This means that physical movement is not a big part of the job. However, drivers still need to be physically fit for this job, as simply sitting for hours at a time can be physically demanding, as anyone with disabilities can tell you.

Drivers will also need to be able to get up and out of the bus several times during their shift. This might be to help load or unload luggage, to speak with a mechanic, to help elderly or disabled passengers enter the bus or for any number of reasons. On the other hand, if there is no need to assist a passenger, or to help load or unload luggage, you might find that your only break from being behind the wheel is at teatime, or during a meal break.

Not all buses have air conditioning and heating, either. These factors vary based on the type of route you will run, the company for which you work and the model of bus that you drive. Therefore, bus drivers may get hot or cold during their daily duties. Not only will you be expected to dress appropriately for the job, but you will also have to accommodate the weather. Temperature extremes can be very uncomfortable for drivers, particularly if the bus company insists on a uniform that is not conducive to comfort during hot weather.

Of course, bus drivers will also have to deal with all manner of different people, from all walks of life. You might find that your passenger load encompasses tourists from the UK and Europe, as well as Americans, Canadians, local Londoners and more.

You will have to be able to deal with all of these travellers, at least on a minor basis. While some passengers might do nothing more than give you a nod as they enter the bus, other passengers will have questions that need to be answered. Still others will try to strike up a conversation with you.

While you do not need to be garrulous to be successful as a bus driver, you do need to be able to be personable, to answer questions to the best of your ability and to maintain your temper if a passenger becomes rude with you.

Let us take a quick look at some of the more important qualities required to become a London Bus Driver. Take a note of them as you may get asked these during your interview.

QUALITIES REQUIRED TO BECOME A LONDON BUS DRIVER

- Customer focused
- Caring
- Considerate
- Physically and mentally fit
- Role model
- Disciplined
- Punctual
- Flexible
- Dedicated
- Be capable of learning routes and procedures

- Knowledge of London and the roads
- Knowledge of the Highway Code
- Knowledge of Health and Safety
- First Aid procedures
- Take a pride in your work and your appearance
- Deal with customers effectively
- Defuse potentially confrontational situations
- An ability to focus and concentrate for long periods of time
- Pleasant
- Conversational
- Have basic numerical and English skills
- Communication skills
- Calm and composed
- Polite
- Courteous
- Safety conscious
- An ability to drive a bus!

Of course, the above list is not exhaustive, but it is certainly a great starting point.

EXPECTED DUTIES OF A BUS DRIVER

Bus drivers have a wide range of expected duties, which can change drastically depending on the type of route that you operate. What sorts of duties will you find here? For instance, you will be responsible for adhering to your route and ensuring that passengers are given a timely travel time (as much as possible with the problems of congestion, accidents on the road and inclement weather thrown into the mix).

If you operate a tour bus, then you will be expected to be able to remain away from home for a long duration, sometimes for a week or more at a time. Tour operators will also be expected to have a higher level of conversational ability and a more easygoing attitude than those drivers operating shorter in-city routes. This is because you will have to spend a considerably longer time with your passengers, so you will have to be able to get along reasonably well with all of them.

If you take overseas bus routes, you might be expected to drive your bus through the Channel Tunnel, or to ride the ferry from Britain to France or another nation. From this point, you might have a route in the destination area that must be completed, before you can make the return trip to London once more.

You will also be expected to alert your passengers to each stop. This is a simple duty, but it is no less important for that simplicity. Some passengers will not know when they reach their destination, so it is up to the bus driver to announce each stop several times, once or twice while approaching the stop and once or twice while stopped for passengers to debark or embark. This ensures that all passengers are able to arrive safely at their destination, even if they are unfamiliar with the area.

At the beginning of each shift, you will be responsible for conducting an inspection of your bus and ensuring that everything is in working order. Anything that seems out of the ordinary should be reported immediately. Any issues relating to safety or roadworthiness should be repaired before the bus leaves for the route.

You will also be expected to take cash for tickets and issue tickets to passengers. For buses with an electronic ticketing system, you will be responsible for ensuring that it is operating properly during your route and for handling cash when the system is not in working order. You might also be responsible for operating a handicap lift for passengers in a wheelchair, if your bus is equipped with such a system.

PERSONAL ATTRIBUTES, INTERESTS AND SKILLS

As you might imagine, being a bus driver in the city of London requires that you have some specific skills, personal qualities and interests. While not all of the items contained in the list below are required, they all help ensure that you are able to complete your job satisfactorily, perform your duties adequately and enjoy the career of being a bus driver.

> Patient, Skilled Driver
Obviously, you will need to have solid driving skills to be a bus driver. You will have to handle a variety of different driving conditions on your route, throughout the year. You need to know how to handle a fully loaded bus in wintry conditions such as snow and ice, as well as in other inclement weather, such as heavy rain, fog and hail. Being a responsible, patient driver is an essential quality and will help ensure that your passengers are able to travel safely from their point of origin to their destination.

> Good Eyesight
Drivers who do not have 20/20 vision are certainly permitted to become bus drivers in the city of London. However, with any vision aids, such as glasses or contacts, you should have 20/20 vision. You need to be able to see clearly at all times, in order to watch the road ahead of the bus, to take

note of your surroundings, to keep an eye on the weather and to ensure safety aboard your bus. Good eyesight is a prerequisite for anyone hoping to become a bus driver.

> Hand-Eye Coordination

Not everyone has good hand-eye coordination. However, it is essential that bus drivers be able to handle their bus to the fullest. This means being able to react instantly when something untoward occurs, such as a car stopped in the road ahead or an auto accident blocking the travel lanes. Sudden occurrences (animals darting in front of the bus or pedestrians crossing the street at the wrong time) can cause serious problems for those who do not have good hand-eye coordination.

> Able to Read Well

Bus drivers certainly have to be literate. There are many different things that have to be read on a regular basis as a bus driver. You will have to be able to read and understand various road signs, as well as reading checklists and other information handed out by the bus operator or the mechanical department. You must also be able to read instructions, comprehend written directions and more.

> Be Able to Handle Cash

Bus drivers are required to handle cash less and less frequently in these modern days. The old system of paying cash to obtain a bus ticket has largely been replaced by an electronic system that issues tickets. However, drivers must still be able to handle cash accurately, to make change for passengers at times and more. If you do not have strong skills with money, the situation can become very difficult. You should be familiar with all common forms of cash, as well as with some foreign forms, and be able to make change for anything that comes your way.

> **Friendly and Approachable**

This is more of a personality trait than a skill, but it is nonetheless important. As a bus driver, you will have to interact with your passengers on a regular basis. You must be friendly and approachable, or your passengers will certainly not have a good trip. They should not feel hesitant to approach you for information or to ask a question. In fact, the reverse should be true. Your passengers should feel welcome to approach you with anything at all. You should be able to carry on a conversation with a stranger, to laugh and joke (appropriately) at times and to answer any question as accurately as possible.

> **Have an Assertive Personality**

There will be times that you have to be assertive on your route. Perhaps you will have a passenger who has had a bit too much to drink. Perhaps you will have a passenger who is already angry from something that occurred before getting on the bus. Therefore, you need to be assertive enough to keep these situations well in hand. You should not be rude, but you must be able to maintain control of your bus. The environment of the bus should be comfortable for everyone and maintaining that environment and level of safety is your responsibility.

> **Calm Under Pressure**

There will certainly be times when you are faced with pressure as a bus driver. Perhaps your bus will suffer a flat tyre. Perhaps an accident ahead of you will create long delays and your passengers will become irate. You might even find yourself faced with a dangerous situation (though these are quite rare). Regardless of the situation that you face, you need to be able to maintain a calm, rational demeanour, to make reasonable decisions based on logic and to keep your head. This is essential.

> Some Familiarity with Other Languages

This is not a hard requirement, but some familiarity with other languages will certainly be a benefit to you as a bus driver. This is particularly true if you will be operating an overseas bus route, as you will often find that you have to communicate with passengers in their own language or to make announcements in a different language.

In the same vein, London bus drivers should also be familiar with the most common local dialects. The city is home to an enormous range of dialects and accents, and a bus driver should be able to understand the vast majority of them. This ensures comfort and safety for both the passengers and the driver.

REQUIRED TRAINING FOR LONDON BUS DRIVERS

Of course, if you are going to be a good candidate for bus driver positions, you need to ensure that you have the right training. The average driving licence is not going to get your foot in the door here. You need to have specific commercial training and the right kind of licence to get started. What do you need?

London bus drivers must have a PCV licence to operate a bus correctly. If you do not have this licence, then you will not be able to apply for a position. PCV stands for Passenger Carrying Vehicle. What should you know about getting this type of licence?

First, you will have to have a basic EU driving permit to get started. Second, you will need to be at least 21 years of age. There are also other requirements for licensing. These include passing medical examinations and meeting health requirements, as well as passing a driving test.

Training for the PCV licence can often be found with the help of the DSA (Driving Standards Agency). This entity is responsible for training all commercial passenger vehicle drivers. The course, itself, lasts a little longer than a full month (about five weeks in total). The training here includes several things:

> Bus driving lessons

These lessons will teach you how to handle a bus in almost any type of driving conditions and will help ensure that you are familiar with the layout of the controls and how the bus operates. You will also learn more about operating the doors and wheelchair lifts here. Driving lessons also familiarise you with how buses handle on the road, which is very different from passenger vehicles or even commercial lorries.

> Induction training

Induction training is another essential ingredient of being a successful bus driver applicant. You will learn many other factors during this training, including how to operate the ticketing system and what to do when the ticketing system is not working. You will gain familiarisation with route structures and how to drive them; how to care for your passengers and how to maintain order; and even more about various health and safety concerns, such as caring for a passenger who is having a heart attack.

Once you have completed your driver training course through the DSA, you are then eligible for a position as a bus driver. However, your test will only result in a basic licence – there are higher grades (Level 2 and Level 3) that can also be obtained (and should be).

The DSA can be contacted with the following information:

Driving Standards Agency
Stanley House
56 Talbot Street
Nottingham
NG1 5GU

Tel: 0115 901 2515
driving-tests.co.uk

PAYING FOR TRAINING COURSES

DSA-authorised training courses are not free of charge. They will have to be paid for. There are several ways in which these courses can be paid for, as well.

> Pay your own way
If you have the money, you can certainly pay for driver training classes out of your own pocket. This is the simplest, most convenient way of going about it, but the classes are not cheap. You will need to pay for the classes upfront. In this scenario, you will not be required to sign a contact with one particular company, nor will you have to repay a sponsor. In addition, bus drivers are in very high demand, so sponsorship is not a requirement for getting a job after completing your training, the way it is in some other industries.

> Company sponsorship
Some bus companies will actually sponsor your training. This is the most affordable way to get your driver training courses out of the way, but it will require that you sign a contract with your sponsor. In this situation, you will work for the sponsoring company for a specified number of years in exchange for them paying for your training. Some companies will not require that you sign a contract, but will instead require that you repay the training fees within a year

of graduating from the course. Some companies might even offer a fee deduction schedule, where the cost is taken from your paycheck over a period of time in small increments.

> Unemployed benefits

If you are currently unemployed, you might be eligible to receive bus driver training at no cost to you. The New Deal Scheme might be able to pay for your training, but you will have to check with your local Jobcentre Plus to determine your eligibility and other details.

INDUSTRY OUTLOOK AND POTENTIAL GROWTH

Now that you know a bit more about how to get the training you need, and the possible job duties that might be expected of you, a few words about the current state of the industry and the potential for growth need to be said. The industry is vast, indeed, with a tremendous current need for skilled, trained, dedicated bus drivers.

Currently, the transportation industry employs more than 100,000 bus drivers throughout the UK. The city of London, itself, has an enormous share of that number. In addition, there are 6,000 different companies from which to choose for employment purposes in the UK. This means that even outside the city of London, there are numerous opportunities for employment, and that number is only expected to increase as the need for transportation and population levels grow.

A London bus driver can expect to start their career earning about £9,000. The top end pay for the industry is about £30,000. However those figures vary considerably based on the company with which you find employment, the route that you drive, the type of bus you drive and even if you are accompanied by a conductor.

For advancement opportunities, bus drivers have several potential options. One of the more common promotions is moving into the control area, where you will coordinate routes, communicate with drivers and more. Another area is inspection, where you would be responsible for safety and mechanical inspections on buses throughout the city. Another potential avenue of advancement is into management of a bus system, or even into providing instruction services to new drivers. Finally, self-employment is an option; bus drivers can obtain their Certificate of Professional Competence in Passenger Transport and begin their own bus or tour company.

CHAPTER 3

UNDERSTANDING THE SELECTION PROCESS FOR BUS DRIVERS

Does everyone who goes through driver training go on to become a bus driver? If not, how does the selection process work? What should you know about the process and how do you ensure that you stand the best chance of being selected to move up to being a probationary driver? This chapter will walk you through the entire process from basic training to being selected as a bus driver.

BASIC TRAINING REQUIREMENTS

As mentioned in the previous chapter, the first step on your path to becoming a fully fledged bus driver is to ensure that you have the basic training offered by a company in compliance with the DSA. You should have an EU driving permit to get started. After graduating from the DSA course, you will have your PCV permit. However, there is more training on the horizon for you.

Basic training requirements ensure that you are familiar with buses and the various systems that are used to operate them. You will also learn more about handling your passengers, handling hazardous conditions and quite a bit more. New drivers will have to undergo a minimum of four hours of theory testing, as well as a full two-hour test in practical application.

The basic testing and study requirements will include the following:

> **A multiple choice theory test for bus and lorry drivers (£35)**
This test ensures that you are familiar with the theory behind driving a bus.

> **Hazard perception test for bus and lorry drivers (£15)**
This test ensures that you are able to accurately identify and react to a variety of hazards that might occur while driving a bus.

> **Practical test for bus and lorry drivers (£115)**
This is the practical test where you put all of your knowledge and training into practice in a real world setting. Some previous tests and training might be conducted on a simulator. However, the practical test will be completed in a real bus in a real driving situation (usually on a training course).

The ultimate goal of your training and qualifying is to earn a NVQ Level 2 permit, in addition to your PCV. The PCV will usually be earned before your NVQ Level 2. What is required for you to earn the NVQ Level 2 permit?

Here is the assessment strategy for the test, which will help ensure that you are able to adequately prepare for this ultimate training goal.

1. Maintain effective working relationships with colleagues

This requirement is relatively self explanatory. You must show that you are able to adequately maintain professional relationship with the people you work with. This means that you must treat all other employees in a cordial manner, and that you show proper respect to them, as well as to authority figures.

2. Contribute to health and safety in the bus and coach work environment

This requirement is a bit more difficult to understand, but it is an important consideration. You must be able to help maintain a safe, healthy work place for yourself and your co-workers. This might mean identifying and reporting safety or health violations, or it might mean something like alerting an authority figure to unsafe work practices or actions on the part of another employee.

3. Provide professional customer service in the bus and coach industry

The bus and coach industry is all about customers. Without customers, there are no jobs. In fact, the bus and coach industry is even more heavily reliant on customers than the average business, because your job will be to transport those customers immediately to their destination. Therefore, it is essential that you are able to show the proper customer service and treat your passengers correctly.

4. Prepare for passenger carrying journeys in a bus or coach

You will need to adequately prepare to carry passengers through a bus route. This might mean inspecting your bus prior to loading passengers, it might mean ensuring that you have adequate supplies on hand for all passengers (such as pillows or blankets for tour buses), or something else all together. The most significant thing here is that it is your responsibility to ensure that you and the bus are adequately prepared for the journey.

5. Provide professional service to disabled passengers on a bus or coach

You will certainly have to transport disabled passengers during your job as a bus driver. You will need to show that you are familiar with how to care for these individuals and what forms of assistance you might be required to show. Examples of this include 'kneeling' the bus and lowering a wheelchair platform. However, personal assistance for passengers might also be required at times.

6. Share information on the operation of the bus or coach

Here, you will need to share information about how the bus or coach operates, how the control systems work, how the bus handles on the road and more. This explanation should show the practical application of theoretical skills, as well as the application of hands-on training that you might have received.

7. Deal effectively with challenging situations and passengers on a bus or coach

Challenges will most certainly arise during your time as a bus driver. You might face hazards on the road, belligerent passengers or both. You might find that a passenger has a weapon, or that the road ahead is blocked by an accident or something else. How will you handle these situations? You will have to explain how you will handle and overcome challenging situations like these to pass the examination.

8. Drive passenger vehicles safely and efficiently

This means much more than simply showing that you can navigate a bus down the road. You will need to demonstrate safety skills when handling a bus. You will also need to show that you know how to drive efficiently, without wasting fuel and that you know how to minimise wear and tear on the bus. This is an important consideration, as bus companies and UK law require that you know how to reduce fuel consumption and limit the environmental impact of your bus.

9. Operate the passenger systems and service

This is an important part of the assessment. You will need to show that you know how all systems operate, including the ticketing system and other systems on the bus (such as air conditioning, heat, lighting, etc.).

10. Deal with emergencies and incidents during a bus or coach journey

As much as you might wish otherwise, there is always the potential for an emergency to arise during the course of your bus travel. This may or may not relate to conditions on the road. However, you will be required to show that you know how to handle a variety of different emergencies.

For instance, what protocol should you follow if your bus is involved in an accident and there are injuries amongst the passengers? What protocol should you follow should one of the passengers suffer a medical emergency? These are only two of the myriad possible emergencies that can arise and that will need to be handled properly.

11. Negotiate and agree tour itineraries with clients

You will need to understand a tour itinerary and you will also need to be able to adequately plan when there are different possible itineraries. You will need to show that you can do this during the assessment.

12. Process fares and receive and match far payments to tickets

You will have to show skill at handling passenger fares and match those fares to the right tickets. Essentially, this boils down to accepting payment, making change and then providing passengers with the tickets they have paid for. However, you will need to show that you can do this adequately in a timely manner, in a real world situation.

13. Manage financial transactions on coach journeys

During lengthier journeys, you will have to conduct several types of financial transactions. You will have to pay for keeping the bus or coach in a parking lot, at the very least. You might have to pay for a hotel room or other requirements, as well. During this portion, you will need to show that you have a full understanding of how these transactions are handled, and how you submit the bill to the bus company.

14. Provide a transport service for disabled passengers

During the assessment phase, you will need to show that you understand how to operate such a service, as well as show such a service in action. This type of situation might arise if you operate a disabled-specific bus service, for instance. This is an essential consideration, as those who are not able to show that they can provide such service will find few positions within the bus industry.

15. Transport accompanied luggage

Your passengers will often have some sort of luggage accompanying them. This is most common with tours, but also with overseas bus routes and longer bus routes, such as night routes. You will need to show that you understand how to safely store your passengers' luggage in your bus, and will physically need to show how this would be accomplished.

16. Transport unaccompanied luggage

There will be times when you are responsible for transporting unaccompanied luggage during your route. This might be simply taking luggage from one bus depot to another, or it might require that you take additional steps to deliver the luggage to its final destination. However, the storage and transport guidelines differ for unaccompanied luggage on buses and you will need to show that you understand how to do this.

17. Operate a school service by bus or coach

Even if you have no interest in shuttling school children on your bus, this is a requirement to get your permit. Because there is always the possibility that you will have school children aboard, you need to show that you have the skills and knowledge required to do this responsibly and safely at all times.

18. Drive passenger carrying vehicles on international journeys

Because the UK is directly connected to mainland Europe via the Chunnel and ferry service, there is a high probability that you will have to provide international service at some point in your career. This might be a temporary thing if the bus company is short on drivers and needs you to fill in for a short time on the international routes, or it might be a conscious career choice on your part.

However, regardless, you will need to show the assessor that you understand the rules, regulations, laws and safety measures that need to be taken when driving a bus across national borders.

ADDITIONAL TESTING INFORMATION

Depending on the training centre through which you go, you might find additional testing and training requirements. Here is a professional development profile released by GoSkills, one of the premiere authorities on training bus and coach drivers. These questions are part of ensuring that you are a well-rounded, educated, conscientious driver.

1. Know the purpose of the gearbox and how to use it to its maximum advantage

You need to understand how different types of gearboxes work, including manual, automatic and semi-automatic gearboxes. You also need to know how the revolution counter operates and why it is important, as well as the basics of fuel consumption.

2. Know how to combine the use of the brake and gearbox to achieve safe, smooth and efficient driving

This applies to braking while driving. Often, you will need to use the gearbox in conjunction with the brakes to ensure a smooth deceleration process. In addition, slowing the bus with both systems helps ensure that you are able to use less fuel. This process is also called "down-shifting" in some circles.

3. Know about fuel consumption

You will need to know how to optimise fuel consumption for buses and coaches. This includes driving habits and methods, as well as braking methods and idling. Fuel consumption is an important consideration when driving a bus, as you are not able to stop for fuel at random points. Refuelling must be kept to a minimum, so you need to know how to make the most of the fuel in your bus' tank.

4. Know how to take good care of your passengers

You should understand the effects of outside stimuli on your passengers, on their comfort and their security. You should also understand how to deal with passengers who have special needs, as well as other types of passengers. You must also understand the effects of your driving on your passengers.

5. Know how to load and unload vehicles and the effects that this can have on handling and performance

This requirement ensures that you know how to properly load and unload your bus or coach. The way in which you load your bus has an immediate effect on how the bus handles on the road. Unloading also has a dramatic effect on the handling of your bus. This applies to loading and unloading passengers, as well as to luggage and other items.

You should understand the load limit per axle of your vehicle, as well as the proper distribution of luggage. Finally, you should understand the tilt test, as well as how loading and unloading your vehicle affects your vehicle's stability.

6. Know how the use the relevant rules governing how you work and how they are administered

You should understand the maximum number of hours that you can work per shift, per EU and domestic law. You should also understand the Working Time Directive, how to read and use tachographs, and the basic requirements of the driver CPC permit.

7. Understand the rules and laws that govern the carrying of passengers

As you are no doubt aware, the laws governing drivers and vehicles that carry passengers differ considerably from those that apply to private vehicles and drivers. You need

to understand the Disability Discrimination Act, the correct use of safety belts within your vehicle, how to use fire extinguishers and more.

You will also need to know and be certified to provide basic first aid, as well as the laws governing the carrying of alcohol on these vehicles. Finally, you will need to understand the laws as they apply to private hire and touring paperwork, as well.

8. Know about the safety record of the bus and coach industry and how it compares to other modes of transport

You need to understand where accidents occur most frequently and what areas are the highest risk for accidents. You also need to understand the underlying costs of accidents, including materials, human cost factors and financial factors, as well.

9. Understand the relevant laws concerning illegal immigrants (primarily for overseas drivers)

You should understand the implications of illegal immigrants for bus drivers, as well as preventative measures to take. You should have a firm understanding of the checklist to use to verify illegal immigrants, as well as an understanding of operator liability and legislation in this area.

10. Understand the risks involved with doing the job of a driver and how to develop techniques that minimise those risks

You will need to have a firm understanding of proper seat adjustment within the vehicle, as it applies to you, the driver. You will also need to understand manual handling of the bus, and the various safety aids available to you and how to use them (alarms, radios, etc.)

11. Know how to look after yourself physically and ensure that you are fit for work

You need to show that you understand the effects of drugs and alcohol on your body and mind, as well as the importance of healthy eating. You also need to show that you recognise the effects of stress and fatigue on your mind and body and how to deal with them.

12. Know how to recognise an emergency and react effectively

What does an emergency look like? How do you react to various emergencies? You will need to understand how to behave in these situations, as well as how to assess a situation and avoid complications in an accident.

You will also need to understand when and how to summon aid, as well as how to assist the wounded and provide basic first aid in these situations. Additionally, you will need to know the expected response from you in the case of a fire, how to evacuate your passengers safely, the expected response from you in the face of aggressive behaviour and how to draft an accident report.

13. Understand how your role as a driver fits within the organisation

You need to understand the roles of others within your organisation and how they relate to your own position, including supervisors, managers and mechanics. You should also understand company standards, as well as sources of company revenue.

14. Have an awareness of the economic importance of the bus and coach industry and how that industry competes with others in the transport segment

You should understand the different types of companies out

there offering transport, as well as the basic requirements for international travel. Finally, make sure that you understand population travel patterns via car, train, aeroplane and other modes of transport.

Developing a firm understanding in all of these areas will ensure that you have the knowledge required to be a successful bus driver and will also affect your potential for advancement. Generally, the better your understanding of these areas, the higher you can move within an organisation and the faster you will be hired by a company.

A NOTE ON ONGOING TRAINING

Many different industries require that professionals undergo periodic ongoing training, and the transportation industry is no different. Beginning in 2008, all professional drivers are required to undergo an addition 35 hours of training every five years. These training hours can be split up over the course of the five-year period, as well. However, there are some hard requirements of which you should be aware.

First, every training period must be at least seven hours in duration. This ensures that all drivers are able to attend a single course per year for the entire five-year period. However, drivers are also allowed to take the training in a single pass, with a full 35 hours of training put in sequentially during consecutive days.

Ongoing training courses for professional drivers are regulated by the government. Therefore, no course or training provider can be offered unless it has been approved by the regulating authority. All courses should offer proof of this approval upfront to drivers. If you find yourself in the position of needing more training, then you should make certain that the course is approved. You can find valuable tips on choosing a training provider through Directgov (www.Direct.gov.uk).

For drivers attending these ongoing training courses, there is no final examination to be passed. You simply need to attend the course and ensure that the training provider has you on the attendance roster for the required number of days. You will also be asked to provide you evaluation of the training course once it has been completed. This helps ensure that all training providers are offering courses that provide real value for those enrolled within them.

BASIC AND ADVANCED QUALIFICATIONS

In essence, your CPC and PCV are all that are required to get started as a bus driver. However, this need not be the end of the road, as far as your education is concerned. For instance, there are numerous other qualifications that can be of benefit to you and help you increase your earning potential. These include the following:

- Level 2 BTEC Certificate, awarded by Edexcel
- Level 2 NVQ (Passenger Carrying Vehicle Driving), awarded by:
  - City & Guilds
  - Edexcel
  - EDI
- Certificate of Professional Competence in National Passenger Transport, awarded by OCR (Level 3 Certificate)

- Professional Competence in International Passenger Transport, awarded by OCR (Level 3 Certificate)

THE SELECTION PROCESS – UNDERSTANDING MITIGATING FACTORS

Of course, not everyone who applies for a bus driving job will be accepted. How will you ensure that you are able to gain employment in this sector? This section deals with mitigating factors in the selection process, and how they affect your likelihood of being accepted.

Test Scores –Your test scores are perhaps the most important mitigating factor here. You will be given several batteries of tests (2 at the minimum). One battery of tests will be for your CPC licence and the other will be for your PCV licence.

If you fail your tests, you will be given a chance to retake them. Usually, you will be given up to three attempts to pass these tests. However, this varies with the testing and training company that you choose to use, or that your sponsor places you with. It is important that you attempt to score as highly as possible on your tests the first go-round. The more times you have to retest, the less favourable your instructors will be of your chances.

Your Work Ethic –Your work ethic is another important factor in being placed as soon as possible. Those who show up for training and testing or to the garage with a good attitude and a willingness to work will certainly be given preference over any applicant who attempts to shirk work or does poorly in completing any assigned duties. Make sure that you are presenting the best face to your supervisors and work hard.

Your Punctuality – As a bus driver, punctuality is essential. Not only will you have to ensure that your route is completed in a timely manner, but you will also have to arrive at the garage early in order to conduct a thorough inspection of your bus before beginning every shift. Therefore, it is more than important that you arrive on time, every time. In fact, if you can arrive about 10 minutes early, it will reflect even better on you. Punctuality is an important factor here.

Cheerfulness/Customer Interaction – As mentioned previously, the job of driving a bus is only partly about being able to handle a large vehicle efficiently. It is also largely concerned with customer service and ensuring that you are able to provide the best, most cheerful face to the public as possible. Therefore, if you are dour and grim at work, you can expect to be passed over in favour of applicants with a more cheerful disposition.

Your Willingness to Learn – Finally, another important mitigating factor is your willingness to learn new things. Those who show an aptitude and willingness to learn and are able to internalise new skills relatively easily, will find that they have considerably more advancement potential than those who are not eager and willing to learn new things. Be willing to learn anything that your instructors and trainers might offer to teach you and your employment prospects will be considerably better.

A FURTHER NOTE ON THE INDUSTRY

An interesting note here is that the bus industry employs a very high percentage of older workers. Numerous workers are nearing retirement age, or at that age already. There are far fewer applicants and new recruits than there are employees within a few short years of leaving the industry. What does this mean for those seeking entrance to being a bus driver? Simply put, it means that there are considerable opportunities for advancement and placement within the bus and coach industry. With the number of soon-to-retire employees, bus companies are seeking to bring in a considerable number of new drivers, as well as other employees, such as mechanics, managers and trainers.

In short, you stand a very good chance of being accepted for the position, so long as you are able to pass your tests and show the minimum requirements for the position. Those with higher aptitudes will certainly see more opportunities for advancement, but the sheer need for new drivers and operators within the bus and coach industry means that there will be more jobs than applicants in a short time. This makes the industry highly attractive and can help ensure that you are able to enjoy a lifelong career should you so choose.

CHAPTER 4
COMPLETING YOUR APPLICATION FORM

As you might guess, getting your foot in the door with a bus company is the most important first step in the process. It all starts with your application form. You need to ensure that you know how to fill out the application completely and accurately. You also need to know what you should expect on the form. This is the key to attracting the attention of your prospective employers. If your application is not sufficiently appealing, you will not be asked to go on to the assessment portion of the hiring process. This chapter will deal with getting your application form completed and submitted, as well as other details in the process.

THE IMPORTANCE OF ACCURACY ON THE APPLICATION

Filling out an application form is an integral part of applying for a job in many different industries. However, it is an especially important aspect of getting on with the right bus company. It is vital that you ensure your application is as accurate as possible.

Every aspect of your application needs to be completed as best you can. However, if you are unable to fill in a particular area, do not put any information in there. This will only hurt your employment prospects in the long run. Accuracy is vital, and you cannot achieve accuracy if you fill in information that is dishonest or incomplete.

THE APPLICATION FORM

It is important to note that the application form you fill out will vary from company to company. However, all companies will ask several very similar questions. This is because most of the same basic information is needed to ensure that you are a good fit. In essence, a qualified candidate with one company will usually meet the requirements of another company.

The following information is a basic outline of an application form, derived from the East London Bus Group. As a note, most other applications follow this guide almost verbatim, with the only differences being slightly more (or less) space in which to detail your previous work experience, your skills and your background information.

EAST LONDON BUS GROUP APPLICATION FOR BUS DRIVERS

The East London Bus Group offers drivers several key informational elements on their website. In fact, the bus driver application can be downloaded directly from the website (www.ELBG.com). Once downloaded, you can simply print the application and then mail it in to this address:

Recruitment Department
East London Bus Group
Head Office
West Ham Garage,
Stephenson Street
London
E16 4SA

Tel: 020 7055 9798

Email: london.recruitment@elbg.com

SAMPLE QUESTIONS:

A wide range of questions are asked during this relatively brief application. The application, itself, is five pages long, and you must ensure that all fields are filled out to the best of your ability.

General Questions about You

- Do you wear glasses?
- If yes, are they tinted?
- Do you wear contact lenses?
- How many years have you held a full British licence? (answer in years and months)
- Provide your full driving licence number:
- If you are under the age of 21, there must be no points on your licence; please confirm that you meet this criteria (check yes or no).
- If you are aged 21 and have held your licence for under two years, there must be no points on your licence; please confirm that you meet this criteria (check yes or no).

> Are there any endorsements on this licence?

> Do you hold a current PCV licence?

> Have you ever held a PCV licence?

Convictions and Legal Proceedings

This section deals with criminal offenses and convictions on your record. The only time a conviction will be taken into account is if it applies to the position that you want (for instance, motoring offences). In addition, you are not required to list any crimes or convictions if they fall under the Rehabilitation of Offenders Act. Contact your Citizens Advice Bureau to determine if your case falls under this heading.

Here, you must list the date, nature of offence and the sentence your offence received. You are given three lines in which to do this. You must then sign and date a declaration stating that you attest the information you have provided is true and accurate.

Application for Employment

Immediately following the listing of criminal offences,. you will come to the personal data section of the employment application. This section also details other aspects of the employment process. You will be required to write in BLOCK CAPITALS for this section, and tick off any boxes to the right if they are applicable. The fields you will need to fill out are as follows:

> The post applied for
In most cases, this will be "driver" or "bus driver".
Make certain that you fill in this field, as the company has no other way of knowing what sort of position you are seeking.

> Your surname

> Your first name

> Your address

> Postal town or area

> Postcode

> Telephone number

> If you are applying for a driving position (or engineering, shunter, or cleaner), you will need to tick the box stating that you are over the age of 18

> Your height measured in feet and inches

> Do you require a work permit to work in the UK? (check yes or no)

> If the answer is no, provide your National Insurance Number:

> How did you hear about this vacancy? (check one or more boxes here; national newspaper, radio, other, local paper, friend or relative)

> Have you previously worked for us or another bus company? (check yes or no)

Previous Employment Section

The next section on the application is a list of your previous employers. This section should date back for the last three years of your work history. Any periods of self-employment, unemployment or assignment to the armed forces should be listed at the top of this section. You are given six lines to fill out.

You will need to include the name and address of the employer, as well as the telephone number where the employer can be reached. Next, you will need to include your start and end dates with that employer, as well as a description of the job duties you performed. In addition, you will need to provide a contact name for reference and your reason for leaving the job. You will also need to indicate whether the company can contact your previous employers for reference.

Why You Are Applying

Now comes the section where you need to tell the company why you want to work with them. Why are you applying? This application gives you a large space in which to detail your reasons for seeking employment with the bus company, as well as outlining any skills or previous experience that might make you a good fit for the company and the position that you seek. Leisure activities, unpaid work and volunteer positions are able to be included here.

Make certain that you make the best case for yourself in this portion. This is an ideal section in which to detail the reasons that you want to be a bus driver. Perhaps you love interacting with other people, or you enjoy being of service to others and providing them with assistance. You should also outline any relevant skills or experience that make you a good fit for the position of bus driver. Remember, this is the place where you are able to impress your potential employer with your desire to be a bus driver, as well as your driving skills.

Let us take a look at a sample response to this question to assist you during your application.

Please give your reasons for applying for htis job and outline andy skills or experience which you feel are relevant to this application. You may include voluntary and unpaid work as well as leisure activities.

I am applying for this job because I have a great interest in providing a high level of customer service, a keen interest in driving and a desire to work in the city of London. I am a caring, dedicated, flexible and safety conscious person who has a proven track record for achievement in previous employed roles.

I have a large amount of experiences that I believe would be suitable to the role and to your company. To begin with, I have 5 years experience where I have been working in a customer focused role. I am totally customer focused and always act as a role model for the company that I am working for. I know how to look after and deal with customers' needs and can also act assertively if the situation arises.

In addition to my customer service skills, I am a competent and excellent driver with over 10 years experience. I have a full driving clean licence and have successfully completed my PCV licence through the Driving Standards Agency. I am a highly safety conscious person who is fully aware of the responsibility that comes with the role of a London Bus Driver. I have a standard Health and Safety qualification and am capable of following company procedures to the letter.

I am physically and mentally fit and attend the gym 3 times per week. I understand that the role requires a high degree of concentration, an expectation which I am capable of meeting. Approximately 2 years ago, I volunteered to work at a local charity shop for 1 hour per week. This allowed me to get a understanding of the community in which I am living and I was able to meet and work with a wide range of people from different backgrounds and cultures.

Finally, in previous company roles, I have successfully passed all training courses and feel fully confident that I would pass any training/on-going training that you company required me to pass. I am dedicated, considerate, caring, flexible punctual and believe that I would be a great asset/ role model to your company.

Equal Opportunities Policy

The section is where the company gathers ethnicity and background information, as well as other specific demographic information. You will need to answer several more questions before finishing this application. These include:

> Your ethnic origins (check the box near the one that best describes you)

> Your age

> Your sex

> Do you have any health problem or disability relevant to your job application (check yes or no)

- If yes, are there special requirements if you have an interview?
- If yes, give details of your condition.
- If yes, could adjustments be made to accommodate you in the workplace?
- If yes, please give details of these adjustments.

While the application form for the East London Bus Group must be downloaded and mailed in, other companies allow you to complete the form online through your web browser. As a note, you will need to have the latest version of Adobe Reader installed on your computer for this.

The Medical Questionnaire

In addition to the general employment application outlined above, you will also need to fill out and submit a medical questionnaire. Of course, if you are accepted for an interview, a full medical exam will also be required, so make sure that you answer the questionnaire as accurately and completely as possible. What questions might you see on such a form? What information does the company need?

Generally, these are very short forms, and you simply check yes or no next to the various questions posed to you. Examples of these questions include the following:

Have you ever suffered from any of these conditions? (Check yes or no)

- Epilepsy
- Psychiatric problems
- Anxiety
- Depression
- Eye disease
- Asthma
- Nervous system disorders
- Heart disease
- Raised blood pressure
- Diabetes

These are only a few examples of the information the company is seeking. At the completion of the form, you will need to sign and date it, attesting that the information you have provided is true and accurate.

As mentioned, you will have to undergo a full medical examination after your application is accepted, so make sure that you provide accurate answers to these questions. Answering "yes" is not an immediate disqualification of your application.

CHAPTER 5
PREPARING FOR THE ASSESSMENT CENTRE

The culmination of your training and the application process should be a visit to the assessment centre. This is where bus companies test potential new drivers on their skills and knowledge. Those who pass the assessment examinations will go on to become trainee drivers. Those who do not will be give the chance to pass the assessment tests again at a later date.

Obviously, this is a very important time for an applicant. How do you ensure that you are prepared? This chapter will outline a basic assessment centre visit and how you should prepare for each segment.

A WORD ABOUT THE ASSESSMENT CENTRE

Assessment centres are essentially training and education centres. A visit to one of these centres will be very different from a one-on-one interview. In fact, the chances are very high that you will be only one of a group of potential employees sent to the assessment centre on the same day.

Most bus companies wait until they have a number of applicants before setting a date for the assessment. This ensures that the company is able to select the best applicants from the group, rather than having to do things more slowly.

One of the benefits of using an assessment centre for such purposes is the fact that you will be with other applicants. This can do a great deal to alleviate stress and tension. It is much easier to be objective and actually enjoy the process when you do not feel singled out for the attention of the test administrators.

There are three main testing areas that you will encounter when you visit the assessment centre. The first is a written test, where you will be tested on your skills in mathematics, your knowledge of English, and your ability to read and to write, an assessment drive with an instructor and, finally, an interview with a manager.

It is vital that you know what to expect in these areas, what the assessors are looking for in applicants and how to prepare yourself for the experience. The following information will provide you with an outline of the process that you can use to your advantage. Towards the end of this section we have provided you with a number of sample test questions to assist you during your preparation.

DOCUMENTS YOU NEED TO BRING

You will need to bring a variety of different paperwork with you to your appointment at the assessment centre. While these requirements will vary from company to company somewhat, you will find that the documents below are "hard" requirements and most companies will insist on seeing them. First, you will need to bring your original driving licence. You will need both the paper and the plastic parts of your licence. You will also need to bring proof that you are legally allowed to work in the UK. For UK citizens, your application with your National Insurance Number and your driving licence are usually sufficient.

For those who are from outside the UK, you will need to provide your work release documentation that states you are able to be legally employed within the country. Usually, this documentation will prove that you have been a legal resident of the UK for at least a six-month period

WRITTEN TEST

The first portion of your assessment centre experience will most likely be the written examination. The written test will cover a number of subjects, but will focus mostly on your command of basic maths skills, as well as your command of the English language and your knowledge of basic road codes.

Maths required to become a bus driver are truly basic. If you did not excel in maths in school, you should have no worries here, so long as you can perform basic functions, such as addition and subtraction. The point of the maths examination is really only to ensure that you will be able to accept fares and make change for your passengers when necessary. There is no need to worry about any intermediate or advanced maths on this examination.

The English portion of the examination is used for a number of purposes. Obviously, the company wants to know that you are fluent in English, as the majority of your passengers will speak this language. This is an essential ingredient. If you are not fluent in spoken English, you will likely not be considered for the position. However, English language skills are also needed for reading and writing. As mentioned in a previous chapter, there will be times that you have to read a variety of different materials, from checklists to instructions to internal memos. You will also have to write on a regular basis, and your writing will have to be legible to others. Of course, you will also need to be able to read road signs and other signage that you will encounter on a daily basis during the course of your duties.

How do you prepare for this examination? If you struggle with basic maths, or even if you feel you have a firm grasp on the subject, it is well worth your time to brush up on your skills. Some simple addition, subtraction and multiplication problems can be valuable refreshers for you in this case. Sample test question will follow later in this chapter.

For the English portion of the examination, you will simply need to know the language. You should also ensure that you have basic literacy proficiency and can read average publications.

Finally, when considering highway codes, your own knowledge and experience is always a good place to start. However, visiting a government website like Directgov (www.Direct.gov.uk) can be a good way to refresh yourself on common codes and learn new ones that you might not have previously encountered. Another website that can be useful here is www.UKHighwayCode.com, where you will find the full UK Highway Code available for viewing or download in a variety of different electronic formats.

THE ASSESSMENT DRIVE

This is the portion of the assessment that most applicants dread. However, it is also the most important part of the assessment where your future career is concerned. You must be able to pass the driving assessment examination successfully. If you are unable to do so, then you will not pass the assessment as a whole and will have to retake the examination at a later date.

While this might sound frightening, most assessment drives only check for basic driving skills. You should be able to perform adequately in these trials regardless of your experience with larger vehicles. In essence, the company wants to see how advanced your skills are and if you will be able to learn others, as well as assessing how well you handle a bus.

The driving assessment will be conducted in an actual bus, and you will be accompanied by an expert assessor/driver. Some companies may give you a brief oral quiz prior to starting the driving assessment, but not all companies do this.

The test actually begins as soon as you enter the bus. You must show that you know common safety procedures, such as putting on your safety belt before starting the bus. You must also listen to what the assessor tells you as you get seated. This is an important part – you will have to listen to the assessor throughout the driving assessment, so start paying attention immediately.

Your test will be conducted in a real bus, but you will be driving on a test course, rather than an actual roadway. Therefore, you should be able to relax somewhat. Test courses are designed to put you through a series of basic driving manoeuvres to ensure that you have at least basic skills behind the wheel. You will also encounter a variety of road conditions during the test. You will also encounter typical "traffic" conditions during your examination.

The instructor will give you directions once you are safely seated. Follow these directions precisely. The instructor will give you directions throughout the duration of the assessment and you will need to follow those directions as closely as possible.

As a note, the instructor will be looking at your overall performance, rather than critiquing you on every single thing that you do. Therefore, if you make a mistake, try not to worry about it. Most mistakes are very minor and can be overlooked. However, the instructor will keep a running tab of your errors. Generally, up to 15 minor mistakes will still yield a passing grade. However, if you make 16 mistakes, you will fail the examination.

Obviously, this gives you tremendous leeway in your driving. As most people will not commit anywhere close to 16 mistakes, you should be able to concentrate on performing the assessment to the best of your abilities, rather than feeling overwhelmed or under pressure that you will not do well. It should be noted that a single "serious or dangerous" mistake will cause you to fail the examination.

As a note, you will be examined on both your general driving abilities and on your performance during set exercises. These exercises will usually involve backing into a garage (executing an S-shaped turn while backing up), as well as a braking exercise and other exercises. Different bus companies will have different set exercises.

How do you adequately prepare yourself for these exercises?

If you have previously completed a training course, then you should be well prepared for this. However, if you are going to attempt to hire on with a bus company that will train applicants, then you need to do things a bit differently.

The best course of action is always to take a class on driving. However, if you are unable to afford the cost of such a course, practising is an essential element of success. Check with friends or family to find someone willing to let you practice with a larger vehicle (the larger, the better).

You should not conduct these practice sessions on the road, though. Find an abandoned parking lot or somewhere similar where you can practice without the worry of causing damage to other vehicles. You might also invest in a set of orange road cones, as these can be valuable aids when learning how to operate a large vehicle such as a bus or lorry.

THE INTERVIEW PROCESS

The interview process is not as stressful as the driving portion of the assessment, but it is no less important that you do well. In fact, this portion of the assessment is so vital that you will find the entire next chapter is dedicated to ensuring that you understand how the process will work, what the interviewer might ask and how to be prepared for the interview.

There now follows two sample tests to assist you during your preparation. There is no time limit during these test, however, there will be during the assessment centre for Bus Drivers.

Please note – the test questions that follow are not the same as the questions you will be required to sit during your assessment. They are for preparation purposes only.

SAMPLE TEST 1

Q1. Which of the following words is the odd one out?

A. Paper
B. Pencil
C. Biro
D. Pen
E. Chalk

Answer [ ]

Q2. Which of the following is the odd one out?

A. Hut
B. Flat
C. Shed
D. House
E. Park

Answer [ ]

Q3. The following sentence has one word missing. Which word makes the best sense of the sentence?

He had spent many years in the same job and was now starting to become ____________ with all of the meetings he was attending.

A. Happy
B. Caring
C. Disillusioned
D. Honourable
E. Frightened

Answer [ ]

Q4. The following sentence has two words missing. Which two words make best sense of the sentence?

The lady ____________ her shopping spree at 9am and ____________ finished at 4pm.

A. started / once
B. started / finally
C. wanted / eventually
D. hurried / wanted
E. commenced / really

Answer [ ]

Q5. In the line below, the word outside of the brackets will only go with three of the words inside the brackets to make longer words. Which ONE word will it NOT go with?

	A	B	C	D
In	(decent	direct	appropriate	abusive)

Answer [ ]

Q6. In the line below, the word outside of the brackets will only go with three of the words inside the brackets to make longer words. Which ONE word will it NOT go with?

	A	B	C	D
In	(coherent	dulgent	believable	candescent)

Answer [ ]

Q7. In the line below, the word outside of the brackets will only go with three of the words inside the brackets to make longer words. Which ONE word will it NOT go with?

	A	B	C	D
In	(doubtable	bred	breeding	cautiously)

Answer ☐

Q8. Which of the following words is the odd one out?

A. Ear
B. Leg
C. Mouth
D. Nostril
E. Eye

Answer ☐

Q9. Which of the following words is the odd one out?

A. Water
B. Lake
C. River
D. Reservoir
E. Pool

Answer ☐

Q10. Which of the following words is the odd one out?

A. Swim
B. Run
C. Sprint
D. Sit
E. Walk

Answer ☐

Q11. Which of the following words is the odd one out?

A. Car
B. Train
C. Trolley
D. Garage
E. Bicycle

Answer ☐

Q12. Which of the following is the odd one out?

A. Milk
B. Tea
C. Coffee
D. Sugar
E. Spoon

Answer ☐

Q13. The following sentence has one word missing. Which word makes the best sense of the sentence?

He wanted a football for Christmas but instead he ________ a rugby ball.

A. Obtained
B. Received
C. Wanted
D. Managed
E. Happy

Answer ☐

Q14. Which two letter word can be placed in front of the following words to make a new word?

Coming Going Shore Line

Answer [ ]

Q15. Which five letter word can be placed in front of the following words to make a new word?

Making Wood Stick Box

Answer [ ]

Q16. The following sentence has one word missing. Which ONE word makes the best sense of the sentence?

The bus driver informed the passengers of the ____________ of delays.

A. Potential
B. Wanting
C. Likelihood
D. Need
E. Lookout

Answer [ ]

Q17. The following sentence has one word missing. Which ONE word makes the best sense of the sentence?

He had served eight years of his sentence when he was ____________.

A. Sentenced
B. Custody
C. Released

D. Rehabilitation
E. Convict

Answer ☐

Q18. The following sentence has two words missing. Which TWO words make the best sense of the sentence?

Marketing strategy is a process that allows an organisation to ____________ its resources on the opportunities that will allow it to ____________ sales and achieve a sustainable competitive advantage.

A. concentrate / direct
B. direct / lose
C. concentrate / increase
D. process / increase
E. deliver / focus

Answer ☐

Q19. The following sentence has one word missing. Which ONE word makes the best sense of the sentence?

A graphical chart is ____________ to provide a visual display of information that would otherwise be presented in a table or text.

A. usual
B. alternatively
C. creation
D. designed
E. usually

Answer ☐

Q20. Which of the following is the odd one out?

A. Bench
B. Stool
C. Chair
D. Sit

Answer [ ]

ANSWERS TO SAMPLE TEST 1

1. A
2. E
3. C
4. B
5. D
6. C
7. A
8. B
9. A
10. D
11. D
12. E
13. B
14. On
15. Match
16. A
17. C
18. C
19. D
20. D

SAMPLE TEST 2

Q1. 37 + ? = 95

A. 85
B. 45
C. 58
D. 57
E. 122

Answer [ ]

Q2. 86 - ? = 32

A. 54
B. 45
C. 108
D. 118
E. 68

Answer [ ]

Q3. ? + 104 = 210

A. 601
B. 314
C. 61
D. 106
E. 110

Answer [ ]

Q4. 109 x ? = 218

A. 1
B. 109
C. 12
D. 10

E. 2

Answer ☐

Q5. 6 + 9 + 15 = 15 x ?

A. 15
B. 2
C. 3
D. 4
E. 5

Answer ☐

Q6. (34 + 13) – 4 = ? + 3

A. 7
B. 47
C. 51
D. 40
E. 37

Answer ☐

Q7. 35 ÷ ? = 10 + 7.5

A. 2
B. 10
C. 4
D. 1
E. 17

Answer ☐

Q8. 7 x ? = 28 x 3

A. 2
B. 3

C. 21
D. 15
E. 12

Answer ☐

Q9. 100 ÷ 4 = 67 - ?

A. 42
B. 24
C. 57
D. 333
E. 2

Answer ☐

Q10. 32 x 9 = 864 ÷ ?

A. 288
B. 3
C. 882
D. 4
E. None of these

Answer ☐

Q11. Following the pattern shown in the number sequence below, what is the missing number?

1 3 9 18 ? 72 144

A. 27
B. 36
C. 49
D. 21
E. 63

Answer ☐

Q12. If you count from 1 to 100, how many 6s will you pass on the way?

A. 10
B. 19
C. 20
D. 11
E. 21

Answer ☐

Q13. 50% of 350 equals?

A. 170
B. 25
C. 175
D. 170
E. 700

Answer ☐

Q14. 75% of 1000 equals?

A. 75
B. 0.75
C. 75000
D. 750
E. 7.5

Answer ☐

Q15. 40% of 40 equals?

A. 160
B. 4
C. 1600
D. 1.6

E. 16

Answer ☐

Q16. 25% of 75 equals?

A. 18
B. 18.75
C. 18.25
D. 25
E. 17.25

Answer ☐

Q17. 15% of 500 equals?

A. 75
B. 50
C. 0.75
D. 0.505
E. 750

Answer ☐

Q18. 5% of 85 equals?

A. 4
B. 80
C. 4.25
D. 0.85
E. 89.25

Answer ☐

Q19. 9876 – 6789 equals?

A. 3078
B. 3085

C. 783
D. 3086
E. 3087

Answer ☐

Q20. 27 x 4 equals?

A. 106
B. 107
C. 108
D. 109
E. 110

Answer ☐

Q21. 96 ÷ 4 equals?

A. 22
B. 23
C. 24
D. 25
E. 26

Answer ☐

Q22. 8765 – 876 equals?

A. 9887
B. 7888
C. 7890
D. 7998
E. 7889

Answer ☐

Q23. 623 + 222 equals?

A. 840
B. 845
C. 740
D. 745
E. 940

Answer ☐

Q24. A rectangle has an area of $24cm^2$. The length of one side is 8cm. What is the perimeter of the rectangle?

A. 22 inches
B. 24cm
C. 18cm
D. 22cm
E. 18 inches

Answer ☐

Q25. A square has a perimeter of 36cm. Its area is $81cm^2$. What is the length of one side?

A. 9cm
B. 18cm
C. 9 metres
D. 18 metres
E. 16cm

Answer ☐

ANSWERS TO SAMPLE NUMERICAL TEST QUESTIONS

1. C
2. A
3. D
4. E
5. B
6. D
7. A
8. E
9. A
10. B
11. B
12. C
13. C
14. D
15. E
16. B
17. A
18. C
19. E
20. C
21. C
22. E
23. B
24. D
25. A

how2become

CHAPTER 6

HOW TO PASS THE LONDON BUS DRIVER INTERVIEW

Interviews are always nervous times for job applicants. Will you make the right impression? What might the interviewer be looking for in an applicant? How do you make sure that you give the right impression? What sort of questions might be asked during an interview?

These are all common questions from those about to undergo an interview. While any job interview can be stressful, the interview for becoming a London bus driver will require that you have some rather specialised knowledge. This chapter will help ensure that you are able to give the best fist impression to the interviewer and get the best results possible.

THE INTERVIEW PROCESS

Once you have undertaken (and passed) your written and

driving assessments, it will be time to move on to the actual interview. This will usually be conducted the same day, within the training centre. Most companies will have you interview with a garage manager. What should you know about the process? What might these managers be looking for in job applicants?

YOUR FRAME OF MIND

It's normal to be nervous before and during an interview. However, you should put your mind somewhat at rest here. The assessment process is designed to do some pretty specific things. The reason that you have are required to take a written test and a driving assessment prior to having your interview is to ensure that you actually have the skills required to do the job. If you have made it to the interview stage, then you are a qualified candidate. You have passed the "weeding-out" portion of the assessment and the manager already knows that you have the skills needed to perform the job adequately. What's the point of the interview then? If the manager already knows that you have the skills needed to do the job, why can you not simply start work as a trainee? While the interview is certainly designed to give the interviewer information about you, it might not be the information that you think it is. What do they seek? What questions will they ask?

Simply put, the interviewer is looking for insight into your character, your personality and your attitude. He or she wants to know more about you, rather than about your skills. He or she wants to know how you will react in certain situations, ones that have little or nothing to do with handling a bus on the road.

ATTITUDE

Your attitude during the interview will certainly be assessed.

Are you open and friendly? Are you able to speak easily with the interviewer? Do you make eye contact during the interview? All of these things are important elements for assessing whether or not your attitude is a good fit for the bus industry. As you will have to work very, very closely with the public, you need to have a very good attitude and a cheerful disposition to be successful.

Your interviewer might ask any number of questions to help highlight your attitude, ranging from questions about how you might deal with difficult passengers to how you might answer incessant questions from passengers. However, questions are only of limited use here. The interviewer will assess your attitude throughout the duration of the interview.

ASSERTIVENESS

Bus drivers have to be assertive, though they also have to be easygoing. In essence, you will need to be easygoing with your passengers and the situations you find yourself within, until you are required to be assertive. Generally, this assertiveness should come into play when safety, security or comfort of your passengers is put in jeopardy. You will also need to be assertive when it comes to adhering to the rules set forth by the government (rules of the road, passenger conduct, etc) and regulations of the bus company.

The interviewer might ask you several questions designed to ferret out information about how you might assert your authority in different situations. Of course, the interviewer will also make an informed decision regarding your assertiveness based on your conduct throughout the entire interview. Therefore, while you should certainly appear open and affable, you need to make sure that you also show assertiveness.

PATIENCE

A bus driver must be a very patient individual. If you are impatient, then you will certainly not find that this is a rewarding career choice. The interviewer will be seeking clues as to how patient you might be. You will have to show that you can be a very patient individual, though assertive when the situation requires it. Why is patience such an important element of the job?

As a bus driver, you will encounter numerous situations that are beyond your control. You might be stuck in traffic, growing more and more behind schedule and dealing with irritated, impatient passengers. You must show patience in such a situation. You might have to deal with a mechanical breakdown of your bus, which will require that you show patience and empathy to your passengers, while communicating effectively with the control hub.

Patience is one of the most important factors that an interviewer will look for while speaking with you. In fact, he or she might engender a situation to try your patience. They might contrive to be called away for a phone call during the middle of the interview, leaving you sitting alone. They might be late to the interview or they might do something else, altogether. Show patience throughout the duration of the interview and you will do far better than those unable to do so.

APPROACHABLE

How approachable are you? You should be open and friendly with all of your passengers. You will need to answer questions from passengers, see to complaints and provide general information to many different people. You need to be approachable for everyone who gets on your bus.

Your interviewer will look for signs of how approachable you might be. One of the most important of these will be the initial meeting with your interviewer. Do you seem friendly and genuinely pleased to see the interviewer? Do you offer a handshake when meeting the interviewer for the first time?

Show that you are friendly, open and approachable and your interview will certainly go more smoothly than otherwise. Answer all questions put to you in an easy, friendly manner and never stint in providing information during the interview.

CALM UNDER PRESSURE

One thing that your interviewer will be looking for above all others is your ability to remain calm under pressure. Driving a bus is not necessarily a high-pressure job, but there can certainly be situations where such pressure is present.

Your interviewer will ask questions designed to test how calm you might be given different scenarios and occurrences. Your answers here, as well as how you answer the questions, will give your interviewer a considerable amount of information and will go into the final decision about your qualifications for the job.

SAMPLE INTERVIEW QUESTIONS

During the interview, you will be asked a series of questions and run through different scenarios. As mentioned earlier, this is all part of the process in determining your qualifications for the position. What questions might be asked? What should your answers be? This section will run you through several potential questions that your interviewer might pose to you, as well as the reason he or she is asking those questions.

Will your previous employers give you a good reference?

This question is designed to do several things. First, it is designed to determine what your job performance might have been like in previous positions. Second, it is designed to help determine your honesty. Remember, the bus company can easily check your references (and they will), so you should give honest feedback here.

If your past employers will give you a positive reference, then you are fine. However, what should you do if your employers will give a negative review? If this is the case, then you need to ensure that you explain to your interviewer why that employer might not give you a good reference.

Attempt to explain it dispassionately, if at all possible. Explain what occurred to create such a negative impression, what you learned from the experience and how you might avoid such a thing in the future.

Tell me a bit about your driving record.

This question is designed to accomplish several goals for the interviewer. Obviously, he or she is looking for information about your driving history, such as your safety record, the number of speeding tickets you might have received, etc. However, this question is also designed to ferret out how honest you might be.

Remember, the bus company is going to check your driving record. This is a requirement, and there is no way around it. Be upfront and honest about your record, citing all relevant items, including speeding tickets, parking tickets and other motoring infractions.

While having some points on your licence is not an immediate "no" to a job, you should make sure that you know the company's guidelines concerning this problem. Most companies want you to have a spotless record – if you do not, then you should probably not apply, unless those incidents are very old or the company allows a specific number of points on your licence.

What are your strengths and weaknesses as a driver?

Here, the interviewer is looking for an honest answer, as well as more information about your character. What are your strong points? These are important, as they are, in effect, your selling points. What makes you a good investment for the company? Why should you be behind the wheel of one of their buses?

You should also make sure that you are honest about any weak areas. Chances are good that those areas were noticed during your driving test or during the written assessment. However, you need to be as honest as possible. You should also inform the interviewer of how you intend to strengthen those areas where you are weakest. This will play a large role in whether or not you are hired by the company at all.

Have you had any motoring accidents in the past?

This question is a bit misleading. Rest assured that the company probably already knows about any accidents in which you have been involved. However, they are looking for an open, honest answer about your past. While accidents will not always be an immediate black mark against you, you do need to ensure that you are upfront about any such incidents in your past.

Why do you want this job?

Remember to read your application form before you attend the interview. Make sure that any responses you provide a consistent with your form. Try to think of something a little bit different, other than simply saying – "It's something I've always wanted to do."

We recommend that you include in your response a desire

to work for a professional organisation, a genuine interest in London, a genuine desire to drive buses and of course, the all important 'customer focus'.

Do you have any experience of working with people from diverse backgrounds or cultures?

Unfortunately, some people feel uncomfortable working with people of different sexes, ages, genders, sexual orientation or cultures. If you fit into this category, don't apply! It is important that you can embrace culture and diversity. After all, you are working in London!

Do you think you will get bored or driving the same route day in, day out?

The obvious answer is no. However, try to provide an example of where you previously carried a repetitive or monotonous task. Try to demonstrate that you enjoyed it.

What are your strengths and what are your weaknesses?

Strengths: Flexible, caring, enthusiastic, dedicated, customer focused and safety conscious.

Weaknesses: A bit of a perfectionist which can sometimes drive my partner mad!

SITUATIONAL QUESTIONS:

You will also be given several situational questions designed to test your skills, your patience and your commitment to customer service. These questions can have a considerable range and you will need to give detailed answers, as well as your reasoning behind any course of action that you might voice. What sorts of situational questions might you encounter?

Question: Your ticketing machinery breaks down in the middle of a transaction. What do you do?

Your interviewer is looking for several key things with this question. One of these is that you know the proper procedures to follow. He or she will also be looking for initiative on your part, as well as how calmly you might handle this situation. While the breakdown of your ticketing system might not seem like a terrible predicament, it can lead to getting behind on your schedule, as well as frustration for your passengers.

Question: You have two passengers that need two £1 fares. They are travelling together, and pay with a single £5 note. What is the correct amount of change to give them?

This question is designed to test your maths skills in a situation where more than one passenger is paying for bus fare out of the same initial bill. You need to be careful in situations like this in real life, as it can be easy to become confused. You need to show that you can accurately calculate the correct change for the fare, as well as handling the pressure from this type of situation.

Question: You are faced with an irate passenger who missed their stop. The passenger was not paying attention, but they are blaming you. What do you do?

This is an excellent example of something that might happen in the real world. While you, as the driver, will certainly do your utmost to ensure that passengers know what stop the bus is arriving at, you cannot force them to pay attention. However, you cannot react with anger here. The interviewer is looking to determine how understanding and empathetic you can be when faced with an angry passenger.

FURTHER SITUATIONAL QUESTIONS:

Give me an example of where you have remained calm in a crisis?

Give me an example of where you have delivered excellent customer service?

Give me an example of where you have carried out a safety-related role?

Give me an example of where you have had to concentrate for long periods of time?

These are only a few examples of questions that you might face during the interview process. However, they show a great deal of what interviewers are looking for during this time. Knowing ahead of time what the interviewer wants from you is the best way to ensure that you are able to land the job that you want.

Most interviewees find that the interview is far simpler than they had expected. Remember, the bus company is looking for reasons to hire you, rather than reasons not to hire you. You simply need to relax, be yourself and remember the information above. You will have an easy time during the interview and will be all set to embark on your career in short order.

HANDLING THE WAITING PERIOD

In many industries, successfully completing the interview usually leads directly to a job. However, the bus and coach industry is a bit different. In fact, most applicants will receive a letter stating that their information has been put on file and they will be notified when a position becomes available. This can be a very big letdown if you are not prepared for it.

In some instances, candidates will be offered a job in short order, usually only a couple of weeks. However, in the vast majority of instances, you will simply be put on a waiting list and told that you will be contacted when a position opens up. How do you handle this interim period? What should you do?

Ideally, the best solution for your needs is to go ahead and apply with another bus company. Your success in getting through the assessment and interview process with one company will lead to much better (and more enjoyable) results with other companies. For instance, if you interviewed and passed the examination with East London Bus Group, you might decide to test with Arriva or another of the major bus companies in London. This does a number of things for you.

First, it ensures that you are not simply waiting around for an answer from the first company. Second, it gives you the ability to get your name and information out there into the industry. The more companies you apply with, the better your job prospects will be. Of course, job prospects in the bus and coach industry are very good anyway.

Finally, applying with other companies gives you the chance to make an informed decision about which company you would like to work with. It can be very easy to think that all bus companies are the same, but they are not. Many things differ from one company to another, including pay, routes operated, shift hours worked, benefits and much, much more. By applying with several different bus companies, you give yourself the best chance to find the perfect company with which to work.

how2become

CHAPTER 7
WHERE TO FIND A JOB AS A LONDON BUS DRIVER

Now that you have learned more about the training process, as well as what to expect in the assessment centre, it's time to talk about job options. As mentioned at the end of the last chapter, applying with more than one Bus Company is certainly advisable. However, where can you start? What companies out there offer you the chance to become a London bus driver? This chapter will give you all the information you need to know about finding companies with which to work.

LONDON BUS COMPANIES

The first step in any career movement is to find a company with which to work. Of course, it helps to know your options in the industry. Within this section, you will find a comprehensive list of bus companies operating routes within the city of London, as well as information about their garage locations, the routes they operate and a bit about this history of each company, as well.

ABELLIO

This bus company operates 40 different routes in total. There are five bus depots involved, and the buses are actually contracted out to London Buses. The company also operates 29 routes throughout the surrounding area (outside of London). All told, Abellio operates more than 500 buses throughout their various service areas. Abellio operates garages in Battersea, Beddington, Walworth, Byfleet, Fulwell and Hayes. The company can be contacted with the information listed below:

301 Camberwell New Road
London
SE5 0TF

Telephone: 020 7788 8550
Email: customer.care@abellio.co.uk

ARRIVA

Arriva is one of the largest bus companies in the city of London, or within the UK for that matter. In total, the company employs almost 50,000 people, and serves one billion customers (throughout Europe). Arriva was founded in Sunderland, and the company still has its headquarters there. Since the company's founding in 1960, it has grown immensely, from a relatively small motor vehicle operation to one of the largest transportation providers within Europe. Currently, Arriva offers services in 12 different European nations, including the UK. The company can be contacted with the following information:

Arriva plc
Admiral Way

Doxford International Business Park
Sunderland
SR3 3XP

Tel: +44 (0)191 520 4000
Fax: +44 (0)191 520 4001
E-mail: enquiries@arriva.co.uk

CT PLUS

CT Plus is a relatively small bus company headquartered in London. HCT Group was founded in 1982 and has experienced considerable growth since that time. Currently, the company offers service in several areas of London, as well as in Leeds, West Yorkshire and in several other areas, as well. While the company is certainly smaller than some other options, it has definitely shown that it has the staying power to find success in this competitive industry. The contact information for CT Plus/HCT Group is as follows:

HCT Group
5th Floor
88 Old Street (Shelter Building)
London
EC1V 9HU

Telephone: 020 7275 2400
Fax: 020 7275 2450
Email info@hctgroup.org

EAST LONDON BUS GROUP

One of the larger bus companies in the UK, East London Bus Group is an excellent choice for job seekers. This company offers routes throughout East London, South East London and along the Thames, as well. In fact, ELBG actually comprises three different entities, or trading names. These are East London Bus and Coach Company Limited, South East London and Kent Bus Company Limited and

East London Bus Limited. As a note, this is the largest independent bus company within the UK, and carries 300 million passengers across the UK each year. You can reach the company with the information below:

Customer Services, East London Bus Group, Head Office,
West Ham Garage, Stephenson Street,
London, E16 4SA

Tel: 020 7055 9600 (0900 - 1700 Mondays to Fridays)
Fax: 020 7055 9762
E-mail: pr.london@elbg.com

For recruitment
Phone: 020 7055 9798
e-mail: london.recruitment@elbg.com

ENSIGNBUS

Ensignbus is the UK's largest dealer of used buses, but they are also a service provider. The company got its start in 1972, and has become the best known of any of the UK's many bus dealers. The company continues to operate several key routes throughout the city of London, and they also operate the Transport Museum, which showcases a vast amount of bus history and historical models, as well. You can contact the company with the information below:

Ensign Bus Company
Juliette Close
Purfleet Industrial Park
Purfleet
Essex RM14 4YF
England

Telephone: +44 (0)1708 86 56 56
Fax: +44 (0)1708 864340

EPSOM BUSES/QUALITY LINE

Epsom Coaches Group began back in 1920, making it one of the more established bus companies in the UK. The company located in Epsom, but grew to offer services throughout the area. Currently, the company has been in business for almost a century, and offers a wide range of different transportation services. The company employs more than 240 people and offers service throughout the city of London and the surrounding area. While Epsom might be one of the smaller bus companies operating in the area, it can certainly be an advantageous choice for applicants. The contact information for the company is as follows:

Epsom Coaches Group
Blenheim Road
Epsom
Surrey
KT19 9AF
United Kingdom

Training: (44) 1372 731712

FIRST GROUP

First Group offers a wide range of transportation services and even has operations in North America. Currently, the company operates 90 routes throughout the city of London and the surrounding area, though many other routes are located throughout the rest of the UK. The company also offers train transportation. The estimated number of passengers carried by First throughout the UK is 250 million. First is also known for its transportation simulator, which is located at their Willesden Junction bus depot. You can contact the company with the information below:

FirstGroup.com

GO-AHEAD

Go-Ahead is one of the larger transport companies in the UK, employing about 25,000 people. In addition, the company serves more than 1 billion customers per year through their various services. Go-Ahead has also earned numerous awards over the years, with 2010 alone bringing in six different awards, ranging from the Sustainable Business of the Year Award to the Information Technology Award. Go-Ahead is an excellent option for aspiring bus drivers, and thanks to their diversity of transport services, drivers can make the move to trains if they prefer. You can contact the company with the following information:

The Go-Ahead Group plc
6th Floor
1 Warwick Row
London SW1E 5ER

Tel: +44 (0) 20 7821 3939

METROLINE

Metroline is one of the best-known companies serving passengers in the city of London. With 250 million passengers per year, 80 bus routes and more than 1,200 buses operating regularly, the company offers numerous employment options. Metroline also employs almost 4,000 people within its business structure. The company's buses travel more than 37 million miles per year (combined), but the fleet age is actually very low, at less than six years per vehicle. You can contact Metroline below:

Hygeia, 66 College Road,
Harrow, Middlesex, HA1 1BE

Tel: 020 8218 8888
Fax: 020 8218 8899
Driving Opportunities: 0800 169 2929

SOUTHDOWN PSV LTD

Southdown PSV Ltd offers bus driving options, bus sales and bus engineering opportunities. The company serves a wide range of areas around the city of London, as well. These include Kent, Surrey, Sussex, Crawley, Horley, Oxted, Westerham, Redhill, Tunbridge Wells and Edenbridge. The company also serves Gatwick Airport. To contact Southdown regarding employment opportunities, use the information listed below:

Recruitment
Southdown PSV
Silverwood
Snow Hill
Copthorne
RH10 3EN

busops@southdownpsv.co.uk

SULLIVAN BUSES

Sullivan is one of the smallest bus companies in London, and they are proud of that fact. The company operates local bus routes throughout a variety of different areas, and offers private hire services, as well as specialising in providing assistance to TV and film crews. Sullivan also offers rail replacement services. To find out more about employment opportunities with Sullivan Buses, use the following information:

Sullivan Bus & Coach Limited,
First Floor, Deards House,
St Albans Road,
South Mimms Service Area, Potters Bar,
Hertfordshire, ENGLAND
EN6 3NE

Telephone:+44 (0) 1707 646803
Fax:+44 (0) 1707 646804
email:admin@sullivanbuses.com

TRANSDEV

Transdev is one of the major bus companies in the capital. The company operates a tremendous number of routes throughout the city, and these include standard buses and coaches, rapid transit buses, trolley buses, light rail, heavy rail and even DRT services. In addition, Transdev's operations reach far beyond the UK. In fact, the company operates in nine different countries, including Australia and Canada, as well as many European countries. Transdev operates 20,000 vehicles, of which 16,000 are standard buses. They also employ almost 50,000 people. To learn more about careers with Transdev, use the following contact information:

Garrick House
Stamford Brook Garage
74 Chiswick High Road
London
W4 1SY

Telephone: +44 (0) 20 8400 6052
Fax: +44 (0) 20 8400 6053
Email: information@transdevplc.co.uk

UNO

UNO bills themselves as "the university bus for everyone". Originally, the company was started to provide service to the University of Hertfordshire, back in 1992. However, the company soon grew to offer bus services throughout London and some of the surrounding area, though they still remain heavily focused on North London. UNO was previously known as UniversityBus, though the name was changed to better reflect the company's goal of being the number one choice for passengers in the city. You can contact UNO with the following information:

Uno, Gypsy Moth Avenue,
Hatfield, Hertfordshire AL10 9BS

Telephone: 01707 255764

OTHER AREAS OF INTEREST FOR BUS DRIVERS

These are the primary bus companies that operate within the various districts of London. However, there are many other companies out there that might make suitable options for you. For instance, you will find a number of different private bus companies that operate throughout London. These include companies within the following industries:

HEALTHCARE

The healthcare industry is enormous. Any number of bus drivers are needed in this sector of the economy. You will find positions available operating buses and coaches specifically for the disabled, as well as for medical transport companies and more. The healthcare industry is currently experiencing phenomenal growth around the world, so this might be a very good segment to enter. With the increasing number of people in need of specialized transport, you can be assured of having a job for life. Many of these buses will actually be minibuses, which are smaller and more easily manoeuvred than their larger counterparts, but the pay is usually about the same.

TOURISM

The UK experiences an incredible level of tourism every year. People come to enjoy the city of London, they come to explore the history of this ancient city. They also come to enjoy the Thames, to visit Bath or to explore Stonehenge. This incredible influx of people annually requires a specialized segment of the transport industry to service it. You can find a vast array of possible employment options working in the tourism industry, from national tours to local, London-wide tours. Many tour companies are small, but there are some larger firms taking advantage of the influx of international travellers every year, as well. This can make for a very rewarding career, particularly if you enjoy meeting people from other nations on a regular basis.

ORGANISATIONS

Numerous businesses and other organisations require the services of skilled bus drivers. While some companies will hire out their transport needs, others experience this need on such a regular basis that it is a better decision to have a transport department in house. You can hire on with one of these companies or organisations and provide bus service to employees for any number of different purposes. Examples include non-profit organisations, as well as some of the larger corporations that call London home.

For those interested in attaining a rewarding career with the potential for advancement and enjoyment for a very long time, few things can be as good as the choice to become a bus driver in the city of London. Buses have an incredible history in this city, with some of the world's most iconic vehicles. In addition, becoming a bus driver in London offers other benefits, including a fast-paced job that is never dull.

Of course, the potential for advancement and diversification is also of considerable interest to job applicants here. Drivers can easily move up to garage management, or they can move into teaching others how to drive. A career in an assessment centre for one of the many bus companies in London can be a very rewarding choice.

For those who want to enjoy this field, becoming a bus driver can be incredibly beneficial. The pay scale can be excellent, the benefits are quite good and employees will find that the industry offers numerous other perks, including the ability to travel throughout the city and meet interesting new people.

Many people have entered the industry thinking that being a bus driver would be a temporary thing, a job held just until they were able to locate something else. A considerable number of those people found the job so enjoyable that rather than leaving within a year or two, they spent decades working as a driver, only to move up within the industry with a trusted transport company.

Becoming a London bus driver is not a terribly difficult process, particularly when you follow the steps outlined within this book. The best course of action for any would-be driver is to find a company that offers training prior to promotion to a full driver position. Of course, that training can be easily paid for out of pocket, meaning that anyone with the interest and drive can find a rewarding career in this field.

Finally, the transport industry continues to grow. This ensures a steady supply of new jobs as bus companies add more routes, add more buses to their existing routes and the number of passengers increases. This promises to provide one of the most stable careers in the world, with truly bright potential for the future.

USEFUL LINKS AND RESOURCES

Throughout this book, numerous links and companies

have been mentioned. This section brings all of those links together into a central location so that you can easily access the resources that you require as quickly as possible. All links have been sorted and organised for easier use.

LONDON BUSES (TRANSPORT FOR LONDON)

London Buses is the organisation responsible for managing the majority of the bus companies that serve the city of London. London Buses manages, mitigates complaints, contracts and organises these companies, but does not provide bus service, itself.

London Buses Customer Services,
4th Floor, Zone G7, Palestra,

197 Blackfriars Road,
London, SE1 8NJ

phone: 0845 300 7000
email: customerservices@tfl-buses.co.uk

TRANSPORT FOR LONDON (TFL)

Transport for London is the organisation that authorizes London Buses. Transport for London also offers maps and detailed guides for getting around the city, as well as bus route information, online ticketing and quite a bit more. For travellers, the site offers guide information about sightseeing in the city. TfL can be found at http://www.TfL.gov.uk.

LONDON BUS COMPANIES

These companies were detailed in a previous chapter, but you will find direct access to the companies' websites from this section. Each of the companies detailed within this section is a bus service provider at the minimum, though many of them also offer rail service and other forms of transport. The largest companies offer training to drivers, engineers, mechanics and other professionals, as well.

- Abellio: www.Abellio.co.uk
- Arriva: www.Arriva.co.uk
- CT Plus: www.HCTGroup.org
- East London Bus Group (ELBG): www.ELBG.com
- Ensignbus: www.Ensignbus.com
- Epsom Bus/Quality Line: www.EpsomCoaches.com
- First Group: www.FirstGroup.com
- Go-Ahead: www.Go-Ahead.com
- Metroline: www.Metroline.co.uk
- Southdown PSV Ltd: www.SouthdownPSV.co.uk
- Sullivan Buses: www.SullivanBuses.com
- Transdev: www.TransdevPLC.co.uk
- UNO: www.UNOBus.info

GOVERNMENT LINKS

The links offer information directly from the UK government, or from sources directly affiliated and sponsored/promoted by the UK government. These links range from driver services to driver training and resources for attaining your CPC and PCV.

THE DSA
www.DSA.gov.uk
The DSA, or Driving Standards Agency, is responsible for ensuring that drivers in the UK understand the rules of the road, safe driving practices and more. The agency also

creates driving tests, which it delivers to more than 400 test centres throughout the UK. In addition, the DSA regulates driving trainers and instructors, and promotes activities for improving driving standards.

DIRECTGOV
www.Direct.gov.uk
This organisation is responsible for helping new drivers and riders learn the rules of the road and the regulations of safe driving. The website offers an enormous amount of information for drivers in all classes, including those who wish to earn their Class D permit (bus drivers). The website also offers information about new laws and regulations, as well as help in finding a viable driver training course.

BUSINESS LINK
www.BusinessLink.gov.uk.
This website offers a vast amount of information about commercial transport, ranging from carrying cargo to passengers. It also offers information about driving for a living, as it applies to lorry drivers, as well as for bus drivers. This resource offers an immense amount of data about drivers' hours and tachographs, as well as driving instructors, the driver CPC and taking buses abroad on the international scene.

THE DEPARTMENT FOR TRANSPORT
www.DFT.gov.uk
This is the definitive authority on transport throughout the UK. This organisation oversees all other agencies and directories in the transport industry. As such, the website can provide comprehensive information for those seeking to gain a better understanding of the transport sector in the UK. However, this website involves all transport, including aviation, rail and shipping, as well as road travel.

GOSKILLS

www.GoSkills.org

While this website is not technically put out by the UK government, this organisation is the premier resource for skills training and career planning. It offers an immense amount of information ideal for those searching for the best options for bus driver training and can even provide detailed career development charts that will help a job seeker make the best possible training decisions. In fact, GoSkills offers training for drivers independent of any transport company operating in the UK, which can be a very sound investment for a job applicant.

JOB WEBSITES WITH BUS DRIVER LISTINGS

The UK is home to a number of comprehensive job websites that grant access to bus driver jobs. You will also find that these sites frequently offer interview guides, training tips, career overviews and other valuable information, in addition to their job listings.

WWW.SIMPLYHIRED.CO.UK

SimplyHired is a global job search website, with a specific UK branch. Job seekers will find more than 800,000 positions listed here, including many in the bus and transportation industry.

WWW.REDGOLDFISH.COM

RedGoldfish.com provides access to many different careers within the UK, including bus driving jobs. The website also provides tools for job seekers, including CV tools, interview tips and tricks, information about assessment centres and much, much more.

WWW.TARGETJOBS.CO.UK

TargetJobs is one of the largest resources for UK job seekers, and offers access to advice, knowledge supplements and other information. Visitors will find access to many different bus driving jobs here, as well as other positions within the transportation industry.

WWW.MONSTER.CO.UK

Monster is a worldwide job website, offering branches in many different areas, including within the UK. In addition to job listings, Monster offers applicants a number of services to help boost their employment prospects, such as CV building tools and more.

WWW.MYJOBSEARCH.COM

MyJobSearch is a job website, but one with a difference. This one aggregates the results from 514 different job websites at once, allowing job seekers to search a multitude of job listings from a single website. The website can also provide information for job hunters, including advice.

WWW.INDEED.CO.UK

Indeed is yet another job website that has a specific branch for UK job seekers. A wide range of different industries can be found with Indeed, including bus driving positions and others within the transportation industry. Indeed also serves Ireland, as well as the UK.

WWW.AGENCYCENTRAL.CO.UK

AgencyCentral is one of the UK's leading recruitment websites and gives applicants access to other recruiting agencies, as well as to job websites. Transportation companies have a definite presence on this website, through recruiting agencies that specialise in these types of services.

WWW.PROSPECTS.AC.UK
Prospects proclaims itself as the "UK's official graduate careers website". The site offers a wide range of career options, as well as courses to further your education and generate better success in job hunting.

WWW.CAREERBUILDER.CO.UK
CareerBuilder allows job applicants to put in their requirements for a position, such as location, pay per year, distance from their home and others, in order to better connect them with the jobs they most want. In addition, the website offers numerous other tools that can help job seekers locate the perfect position for their needs.

DRIVER TRAINING CENTRES IN THE UK

These centres offer drivers the training required to jumpstart their careers. These are important elements for those job applicants who do not wish to sign on with a bus company that offers training through their own programs.

DRIVER TRAINING CENTRE
www.DriverTrainingCentre.co.uk
This company offers training and testing for those who want to become bus drivers, through their in-depth courses. They also offer information about studying for theory and hazard perception tests, as well as many other important subjects that soon-to-be drivers need to know.

RITCHIES TRAINING CENTRE
www.RitchiesTraining.co.uk.
Ritchies Training offers drivers access to the courses they require to become licensed bus drivers. This company

has a considerable presence in the training industry, and is frequently used by transport companies for their own training needs. In addition, Ritchies offers classes for individuals wishing to study and become licensed on their own.

TAUNTON TRAINING CENTRE:
www.TauntonTraining.co.uk.
Taunton Training provides driver training for lorry drivers, as well as for those attempting to become bus drivers. The training centre offers a full range of courses for students, as well as offering vital assistance with licensing requirements and applications. Taunton is also a certified ongoing training centre for drivers with their certification already.

TOCKWITH TRAINING CENTRE
www.TockwithTraining.co.uk.
Tockwith Training is a fully licensed training facility for commercial drivers in all manner of industries. Tockwith can train drivers for LGV, HGV, PCV and any other types of driving. This company is also an approved provider for Periodic Driver CPC training for those drivers who need to earn their 35 hours toward recertification.

Please note: How2become Ltd is not acting in conjunction with any companies/organisations contained within this guide.

Visit www.how2become.co.uk to find more titles and courses that will help you to pass the London Bus Driver selection process, including:

- Numerical Reasoning and Verbal reasoning testing resources.
- 1-Day courses.
- Interview books and DVD's
- Psychometric testing books and CD's.

www.how2become.co.uk